YOUNG SCIENTIST

D1369259

Volume 4

The living world

World Book International

World Book, Inc.
a Scott Fetzer company
London Chicago Sydney Toronto

Illustrated by : Maggie Brand
David Cook
Sheila Galbraith
Jeremy Gower
Annabel Milne
Jeremy Pyke
Gwen Tourret
Pat Tourret

Acknowledgements

The publishers of **Young Scientist** acknowledge the following photographers, publishers, agencies and corporations for photographs used in this volume.

Cover	CNRI (Science Photo Library)
6/7	Biology Media (Science Photo Library); Norman Myers (Bruce Coleman Ltd)
8/9	Jane Burton (Bruce Coleman Ltd)
16/17	Eric Grave (Science Photo Library)
20/21	Michael Abbey (Science Photo Library)
22/23	Dr Tony Brain and David Parker (Science Photo Library)
24/25	Alfred Pasieka (Bruce Coleman Ltd)
30/31	Avril Ramage (Oxford Scientific Films)
32/33	Reinhard (ZEFA Picture Library)
34/35	Michael Chinery; Bruce Coleman Ltd
36/37	Peter Ward (Bruce Coleman Ltd)
38/39	Jeff Simon (Bruce Coleman Ltd)
40/41	A. Wharton (Frank Lane Picture Agency Ltd)
50/51	Jane Burton (Bruce Coleman Ltd); CNRI (Science Photo Library)
52/53	Kim Taylor (Bruce Coleman Ltd); Eric Crichton (Bruce Coleman Ltd)
54/55	Kim Taylor (Bruce Coleman Ltd)
56/57	John Cancalosi (Bruce Coleman Ltd); Sinclair Stammers (Science Photo Library)

Published by
World Book, Inc.
525 West Monroe Street
Chicago, IL 60661
U.S.A.

ISBN 0-7166-6303-1

Printed in the United States of America

6 7 8 9 10 99 98 97 96 95

Cover photograph
Large numbers of bacteria are found everywhere. Many of them are useful, but some spread disease. The bacterium in this picture can infect humans as well as some farm animals.

Contents

Giant sequoia trees can grow to be more than 80 metres high. Many giant sequoias are several thousand years old.

All these are living things

How does life begin? All life is made from tiny living things called **cells**. All the different kinds of plant and animal in the world are made of cells. Some cells live on their own, as tiny plants and creatures. Most of these plants and creatures are so small that you need a microscope to see them. Other cells build larger living things by growing and dividing, but remaining attached. There are millions and millions of cells in your body.

All shapes and sizes

The largest living things in the world are trees! Some giant sequoia trees grow in the United States of America. The sequoia can be as tall as 80 metres and can weigh more than 2,000 tonnes. That's about as high as a 30-storey building and heavier than 20 blue whales.

A sequoia tree will probably contain millions of cells. But an amoeba is made up of only one cell. There are all kinds of living things between the simple, one-celled amoeba and the giant, more complex sequoia.

You usually need a microscope to see an amoeba. An amoeba is a tiny living thing that consists of only one cell.

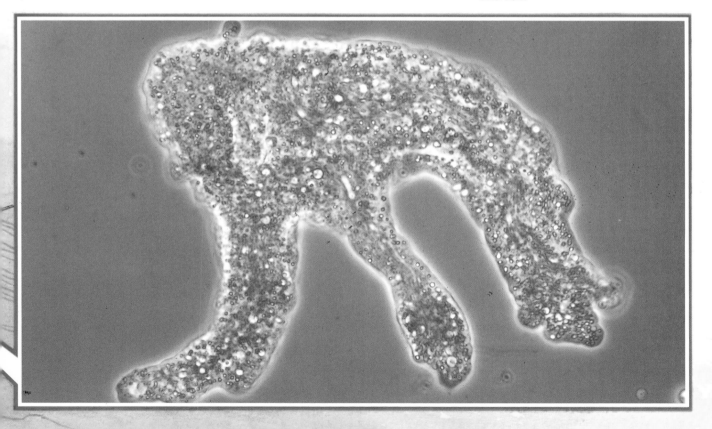

Swans and hares are living things that eat, reproduce and respond to the world about them.

Look at life

Of all the planets travelling around the Sun, only the Earth has living things on it. As far as we know, there is no other life anywhere else in the universe.

Why is there life here and nowhere else? One reason is that the Earth is the only planet with plenty of **water** on its surface. Nearly three-quarters of the Earth's surface is covered with water in the form of oceans, ice caps, lakes and rivers. All life on Earth depends on water. In fact, every living thing is made mostly of water. Your body is about two-thirds water. And a tomato is nearly all water!

Another reason why there is life on Earth is that this planet has a supply of a gas called **oxygen**. Almost all living things need oxygen, which is part of the mixture of gases in the atmosphere all around the Earth.

Not all things on Earth are living things. Scientists who study living things can tell whether an object is living or non-living by asking these simple questions.

Does it eat?

All living things have ways of taking in the substances they need to grow and survive.

Does it respond to the world around it?

Living things **respond** to light, sound or touch. They all have some way of receiving information about the world around them, such as seeing, hearing, smelling or sensing.

Does it reproduce?

All living things make copies of themselves. When they die there will still be creatures like them alive. Making copies is called **reproduction**.

When an insect lands on a sundew plant, the plant responds to the touch of the insect. It traps it with the sticky liquid on its leaves. The hairs on its leaves curl around the insect and hold it. The plant absorbs nutrients from the trapped insect.

Dividing things into groups

We can arrange everything in the world into three groups. Some things are living now, others were once living and some were never living.

A table is not living. Tables don't eat and grow, or produce young tables! And they don't have eyes and ears or other ways of sensing the world around them.

But if a table is made of wood, then it was once part of a growing tree. The table is not a living thing, but the wood was once part of a living thing.

If a table is made of stone, then it was once part of a large rock. Rocks are not living. They have no senses, and they don't eat or reproduce. So a stone table was never living.

Dividing things into groups helps us to understand how all the different things in the world fit into a pattern.

Plants

club moss

beech tree

tongue fern

rose

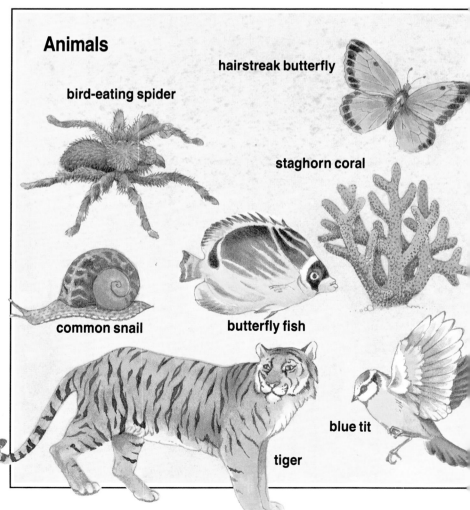

Animals

bird-eating spider

hairstreak butterfly

staghorn coral

common snail

butterfly fish

blue tit

tiger

Find out more by looking at pages **14–15**

The five kingdoms

Scientists who study living things are called biologists. Biologists divide things into different groups. This is called **classifying**.

Some biologists classify all living things by dividing them into five large groups called **kingdoms**. The two main kingdoms are those of **animals** and **plants**.

How can you tell the difference between an animal and a plant? Most animals move around, but most plants are rooted in one place. All animals eat plants or other animals, but plants have a way of making their own food. Sunlight helps green plants to make food from water and a gas called carbon dioxide.

It's quite easy to tell that a tiger is an animal and that a tree is a plant. But what about coral? Coral grows in the sea, and looks like a plant. But it is made up of millions of tiny animals and eats smaller swimming animals.

Is a mushroom a plant? Mushrooms and other fungi don't make their own food helped by sunlight. Instead they feed off living and rotting plants. But they're not animals! So some biologists say that **fungi** belong to a separate kingdom of their own.

There are other kinds of living thing that are different from plants and animals. Most of them are so small that you can't see them without a microscope. They are divided into two more kingdoms. One is made up of creatures called **protists**, which include the amoebas. The other kingdom is made up of tiny creatures called **monera**, which are mostly bacteria.

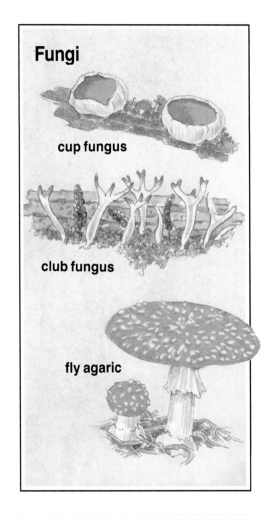

Fungi

cup fungus

club fungus

fly agaric

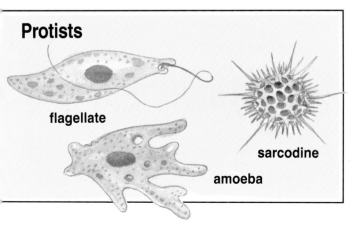

Protists

flagellate

sarcodine

amoeba

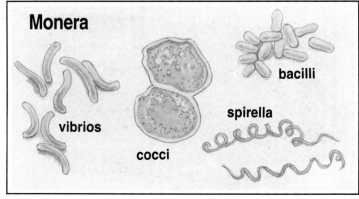

Monera

bacilli

spirella

vibrios

cocci

What are these living things?

There are more than two million different kinds of living thing in the world. On pages 10 and 11, you saw how biologists have classified them into five kingdoms. Can you tell which of the living things on these pages are animals, plants, fungi, protists or monera?

Divide a clean sheet of paper or a page in your notebook into five columns, one for each kingdom. Now write the name of each of these living things in the correct column.

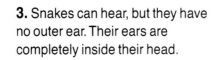

Animals	Plants	Fungi	Protists	Monera

1. A female tiger brings meat for her cubs to eat.

2. Owls hunt at night.

3. Snakes can hear, but they have no outer ear. Their ears are completely inside their head.

Find out more by looking at
pages **10–11**
14–15

4. This spider is spinning its web.

5. The sea anemone eats tiny creatures in the water.

8. Amoebas live in ponds, moist places and in the bodies of animals and human beings.

7. These cocci are a kind of bacteria that can cause diseases.

6. The leaves of this convolvulus plant are turned towards sunlight.

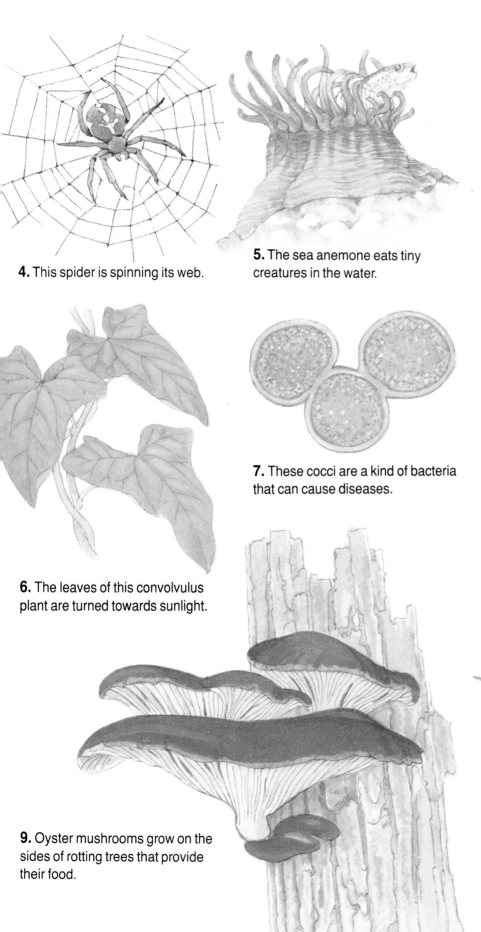

9. Oyster mushrooms grow on the sides of rotting trees that provide their food.

10. Pine trees produce seeds in female cones.

From kingdom to species

Each of the five kingdoms is a huge group. Scientists already know about more than one million different kinds of animal in the animal kingdom, and about more than 350,000 kinds of plant in the plant kingdom. Every year, hundreds more kinds are being discovered, and every year, hundreds are dying out, or becoming extinct.

It's very difficult to think about such an enormous number of plants and animals all at once. So scientists divide each kingdom into smaller groups of living things. These have particular features in common.

Classifying tigers

Each kingdom in the living world is divided into six smaller groups with different names. Scientists call these groups **species**, **genus**, **family**, **order**, **class** and **phylum**. The smallest of these groups is the species. Let's see how the tiger fits into the different groups.

The largest group is the phylum. There are more than 20 phyla in the animal kingdom. The plant kingdom has 10 phyla. All the different kinds of animal or plant in a phylum have only one feature in common. There is only one kind of animal or plant in each species, but there may be hundreds or even thousands of species in a phylum. Each species of animal or plant can produce copies of itself.

Species

Every tiger is an individual creature, but all tigers belong to the same species of animal.

tiger

Genus

Different species that are very similar, such as tigers, lions and leopards, are part of a group called a genus.

tiger

Family

Living things from more than one genus can have common features that make them part of the same family. Tigers are part of the cat family.

tiger

Order

Wild cats and bears both eat meat. Meat-eating animals make up an order, or group, called **carnivores**.

tiger

Class

Sheep and rabbits don't eat meat like cats and dogs. But all these animals feed their young on milk. They all belong to the class of **mammals**.

tiger

Phylum

All the animals in this group, or phylum, have something in common. They all have a backbone. They belong to the phylum of **vertebrates**.

tiger

Make a model of an animal cell

You will need:

water

a small, solid, plastic ball

a small plastic bag

scissors

some gelatine

1. Ask an adult to help you dissolve the gelatine in some hot water. Add enough warm water to make a jelly that will set. Leave the jelly to cool for a few minutes.

2. Pour half of the jelly into the plastic bag. Keep the rest of the jelly warm, while the jelly in the plastic bag sets. It will set more quickly in a cool place.

3. When the jelly in the bag is partly set, put in the plastic ball. Then add the rest of the jelly. Tie a knot in the bag to seal it and trim off the ends.

This is roughly the shape of an animal cell. The plastic bag is like the membrane. The plastic ball is like the nucleus, and the jelly represents the cytoplasm.

The smallest living thing

Do you know what your body is made of? It's made of skin, bone, blood, muscles and fat. But what are these made of? All the different parts of your body are made up of tiny living things called **cells**. All plants and animals and other living things are made up of cells. Some very small forms of life are made of just one cell. But in the human body there are more than 10 million million cells. Cells are the building blocks of life. All living things are made up from at least one cell.

This is a diagram of an animal cell. Animal cells of different shapes and sizes carry out different jobs.

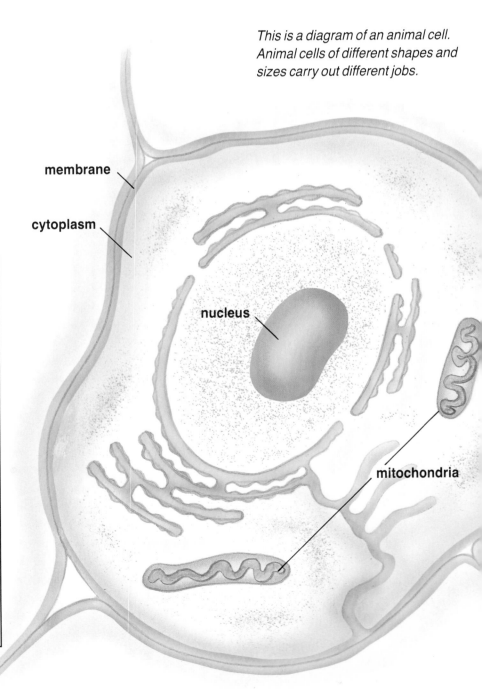

membrane

cytoplasm

nucleus

mitochondria

What's inside a cell?

Inside a cell is a jelly-like substance called **cytoplasm**. This is contained inside a kind of skin called a **membrane**. Plant cells also have another skin called a **cell wall**, outside the cell membrane. The cytoplasm contains many different working parts and a large number of different chemicals, which all have a job to do in the cell.

The most important part of the cell is the **nucleus**. The nucleus is the control centre of the cell. Without a nucleus the cell would die. Other important working parts of the cell's cytoplasm are called **mitochondria**. They provide almost all the energy the cell needs.

Chemical substances called proteins exist in every cell. They are essential to plant and animal life. Some of these proteins are called **enzymes**. An enzyme brings chemicals together and helps them to react. The chemical reactions which enzymes are involved in are vital to the survival of every cell.

Find out more by looking at pages **18–19**

The paramecium is a tiny living thing with only a single cell. It is found in ponds. The paramecium cannot be seen without a microscope.

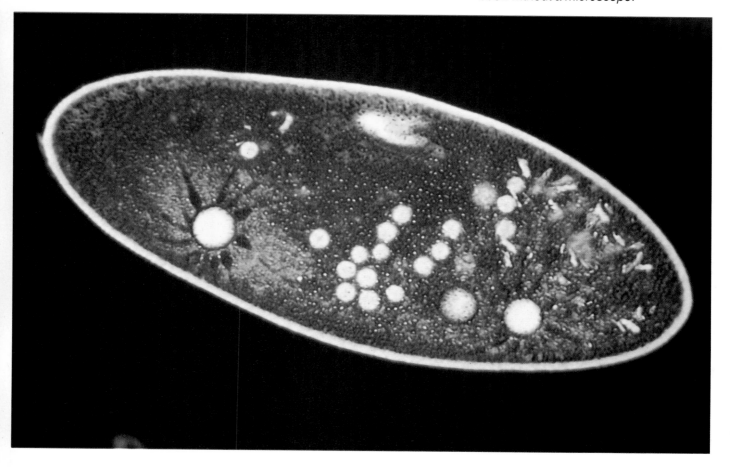

18

Find out more by looking at
pages **16–17**
      34–35
      40–41

Life under the microscope

Most cells are so small that you can only see them through a microscope. They have to be magnified about 2,000 times before they can be seen. There are thousands of cells in something as small as a poppy seed. So you can imagine how very tiny cells can be. But you can see some single cells, such as the yolk of a bird's egg, without a microscope. The largest cells are the yolks of ostrich eggs, which are about the size of a football.

Cells are different sizes and shapes depending on the job they do. Many single-celled plants and animals look like tiny balls or boxes. The cells in larger plants are mostly cube-shaped or oblong-shaped. Muscle cells are long and thin so that they can stretch and shrink when an animal moves. Nerve cells have branches that reach out in many directions. Some help animals to feel different sensations.

Cells are living things that feed, reproduce and respond to the world around them.

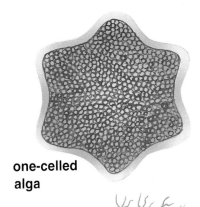

**one-celled
alga**

nerve cell

Making new cells

Cells make new cells by dividing into two or more parts. Some kinds of cell, including single-celled animals and plants, split into two parts that are both exactly the same.

Here you can see a plant cell dividing. The cell has a larger nucleus when it is about to divide. The nucleus starts and controls the dividing process.

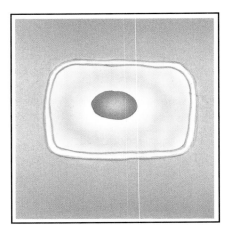

1. The nucleus of the cell is ready to divide.

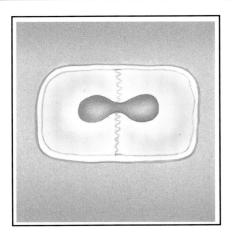

2. The nucleus has almost split in two. A new layer begins to form dividing the cells.

How does a cell eat?

Cells don't have a mouth to eat with. Animal cells take in the substances they need through their outer skin, or the cell membrane. This membrane is made up of chemicals that let useful substances into the cell, and keep unwanted ones out.

Plant cells contain small working parts called **chloroplasts**. These contain a green substance called **chlorophyll**, which makes the plant's leaves green. Chlorophyll takes light energy from the Sun, which helps the plant to make its food.

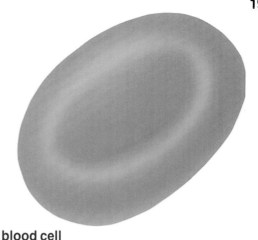

blood cell

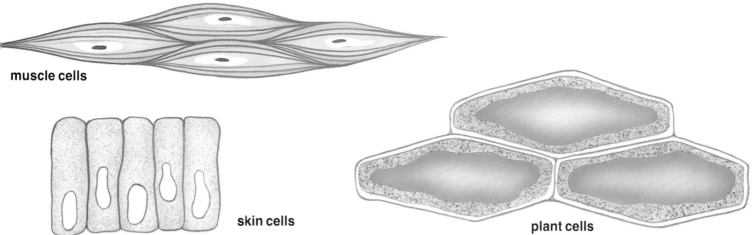

muscle cells

skin cells

plant cells

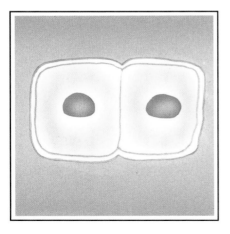

3. The nucleus has now divided into two separate parts.

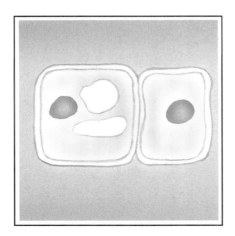

4. The cytoplasm builds up a wall around the new dividing layers.

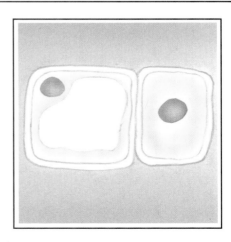

5. The cell has now divided into two separate cells. One cell is ready to divide again.

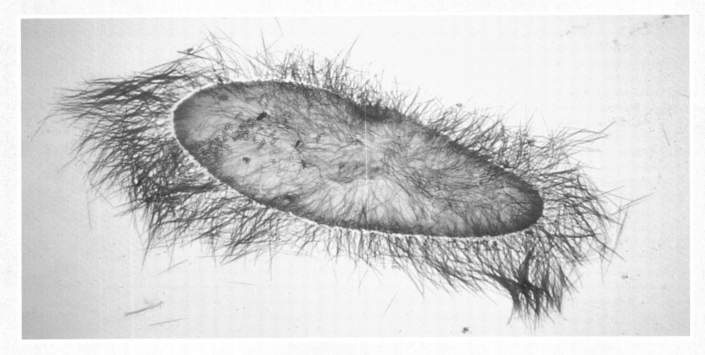

This paramecium is covered with hairs that help it to move along. The hairs also catch particles of food.

What is a protist?

Protists are small living things that can't easily be classified as animals or plants, fungi or monera. The kingdom of protists contains all the different species that don't fit into the other four kingdoms.

Hairy protists

Under the microscope, some protists look as if they have a hairy coat. Their cells are covered by hair-like threads. These 'hairs' enable the cell to move about. By moving all together, they push the cell along. Biologists call these threads **cilia**, and name the creatures **ciliates**.

One kind of ciliate is called a paramecium. It is only a single cell, but it has a kind of mouth and stomach to eat and digest its food.

What are euglenas?

Euglenas are one of the strangest kinds of protist. Euglenas can make food using sunlight, as plants can. But in other ways they are not like plants at all. Euglenas can move around, using a long, whip-like projection called a **flagellum** that they beat backwards and forwards.

The euglena uses its flagellum as a kind of oar to move itself around.

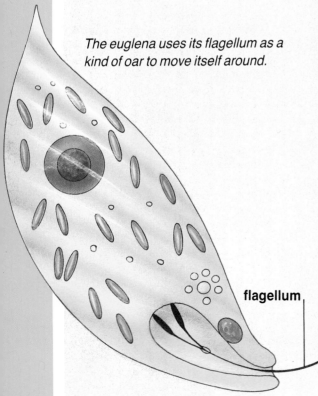

flagellum

Amoebas

Amoebas are single-celled protists. Giant amoebas are just big enough for you to see without a microscope. Amoebas eat their food by wrapping themselves around it and then taking it in, through their outer cell membrane.

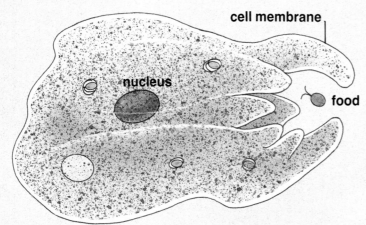

A single-celled amoeba moves about in the water until it finds food. Part of the amoeba's body surrounds the food, which is then absorbed through the membrane.

The food floats inside the amoeba's body until it is absorbed. Then any waste can pass out through the membrane.

How does an amoeba move?

You can use the model cell described on pages 16 and 17 to show how an amoeba moves.

1. Hold the model steady with one hand and pull one side of the plastic bag out with your other hand.

2. Now squeeze the jelly into the part of the bag you have pulled out.

3. Repeat this several times, and you will see that your model has moved forwards a little way.

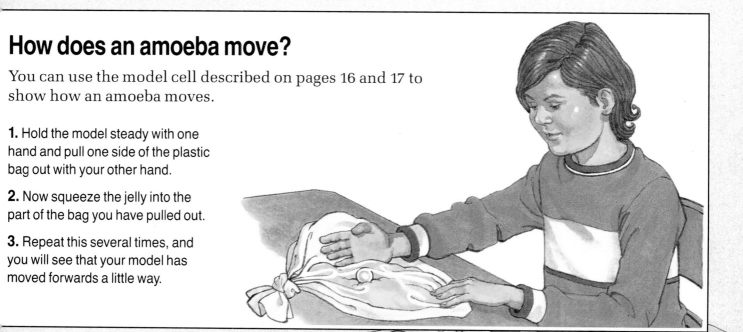

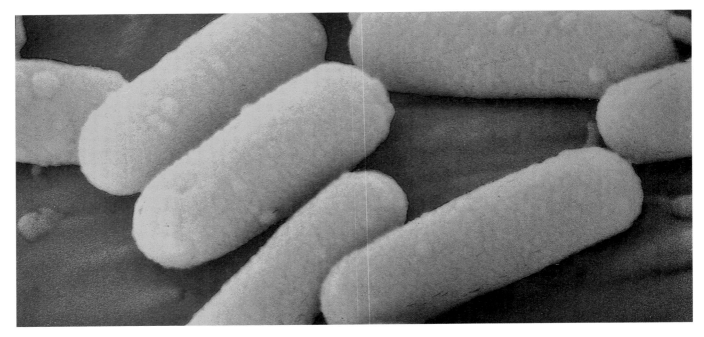

These rod-shaped bacteria are on the head of a pin. The photograph is enlarged so much that the surface of the pin looks rough, not smooth.

What are bacteria?

Bacteria are very small living things that have only one cell. Some scientists classify bacteria as plants, because some kinds of bacteria can make their own food from sunlight, like green plants do. But there are many differences between plants and bacteria, and so most scientists say that bacteria belong to a separate kingdom of living things, which is called the **monera** kingdom. There are thousands of different kinds of bacteria. Each bacterium is so small that you need a microscope to see it.

Where do bacteria live?

Some kinds of bacteria live on you! There are millions of bacteria on your skin, and inside your mouth, nose and lungs. A great many live in your intestines. Other kinds of bacteria live in the air, in water or in the top layers of soil.

Harmful bacteria

Bacteria can be spherical, rod-shaped or spiral. Each of these harmful bacteria causes a serious illness or disease.

Spherical bacteria

pneumonia

**boils
blood poisoning**

**sore throat
scarlet fever**

Find out more by looking at pages **18–19** **30–31**

What do bacteria do?

Many kinds of bacteria are very important and useful. They help living things to survive and be healthy. Some of the bacteria on your skin protect you from other tiny living things that might harm you. Those in your intestines help to break down the waste products that your body doesn't want. They even help to make vitamins that keep you healthy. We use bacteria to make some types of cheese and yoghurt.

Bacteria in soil and water help to break down animals' droppings and the dead bodies of animals and plants. These all contain chemical **elements**, such as carbon and nitrogen. Some bacteria help change chemical substances such as nitrates. These can then be used again by other living things.

But not all bacteria are helpful. Some kinds of bacteria destroy healthy cells and cause diseases. Whooping cough and food poisoning are caused by bacteria. Animals and plants suffer illnesses, too. Anthrax, a cattle disease, is caused by bacteria. Bacteria also cause certain kinds of rot and blight in plants.

How do bacteria eat?

Bacteria are surrounded by a thick cell membrane with no openings to take in food. Some bacteria contain chlorophyll, which can make food with the help of sunlight. Others soak up fluids from the body they live in.

Bacteria that live on rotting plants or animals use proteins, called **enzymes**. These can turn the dead matter into simple liquids. The bacteria then soak up the liquid through their cell membranes.

In this rod-shaped bacterium, the nucleus is surrounded by cytoplasm.

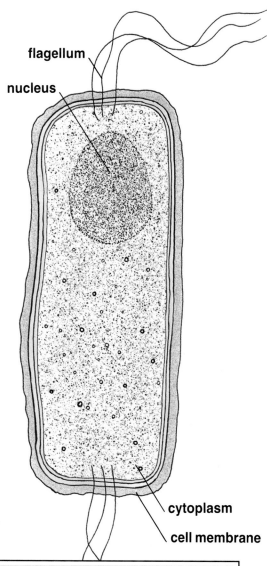

flagellum

nucleus

cytoplasm

cell membrane

Rod-shaped bacteria

tuberculosis

tetanus

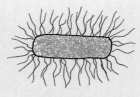

typhoid fever

Spiral bacterium

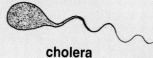

cholera

What is a virus?

Viruses are the smallest and simplest living things. You can only see them through a microscope called an **electron microscope**. Viruses are made of chemicals which are just like those in our own bodies. Some scientists say that viruses are not really alive. They can only survive by entering the cells of animals, plants or bacteria. And viruses can only make copies of themselves with the help of other living cells.

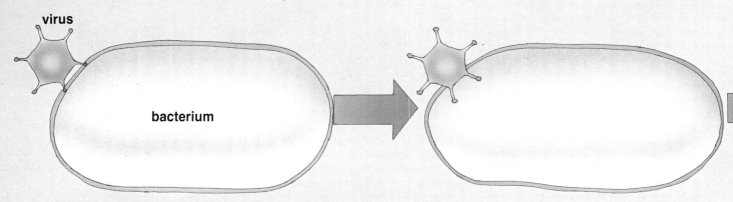

virus

bacterium

1. This virus has attached itself to a bacterium.

2. The virus has pierced the cell wall of the bacterium.

Living cells can be attacked by a particular virus. This influenza virus can make you feel ill.

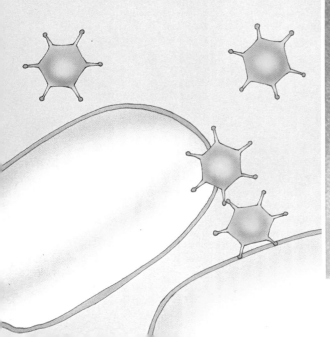

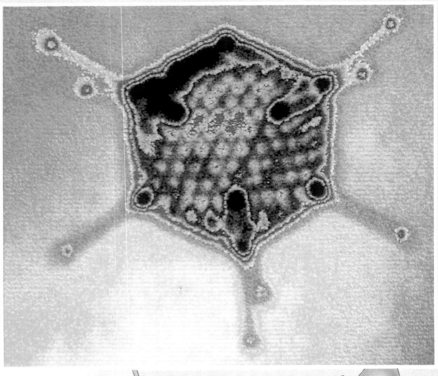

Taking over animal cells

By itself, a virus is a lifeless particle which cannot reproduce.
So it needs to be carried into a living thing in some way.
Viruses often pass into an animal's body when the animal
breathes. Once inside a cell, the virus uses the cell's materials
to live and reproduce. The virus can make hundreds of copies
of itself. The healthy cells are taken over and destroyed,
which could make the animal ill.

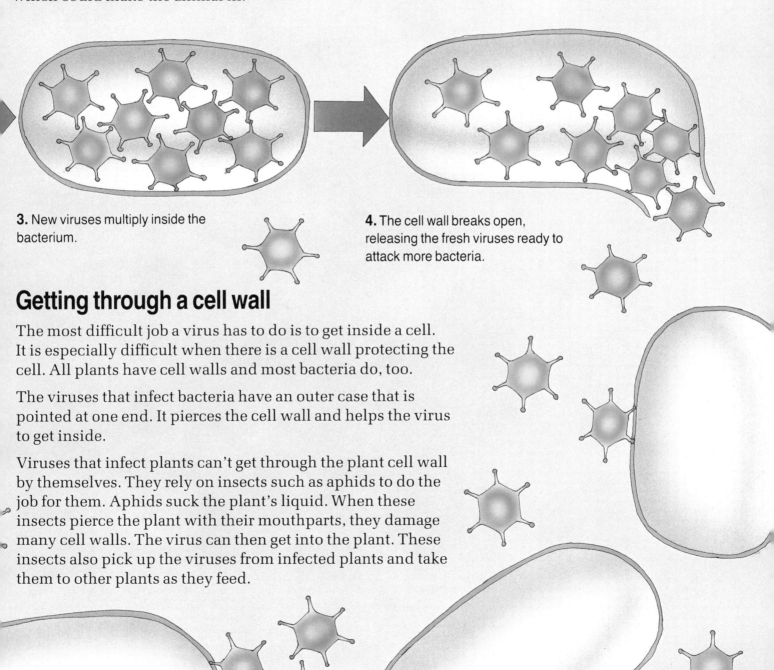

3. New viruses multiply inside the bacterium.

4. The cell wall breaks open, releasing the fresh viruses ready to attack more bacteria.

Getting through a cell wall

The most difficult job a virus has to do is to get inside a cell.
It is especially difficult when there is a cell wall protecting the
cell. All plants have cell walls and most bacteria do, too.

The viruses that infect bacteria have an outer case that is
pointed at one end. It pierces the cell wall and helps the virus
to get inside.

Viruses that infect plants can't get through the plant cell wall
by themselves. They rely on insects such as aphids to do the
job for them. Aphids suck the plant's liquid. When these
insects pierce the plant with their mouthparts, they damage
many cell walls. The virus can then get into the plant. These
insects also pick up the viruses from infected plants and take
them to other plants as they feed.

Food chains on the land

What did you have to eat today? Did you have vegetables, fruit, cereal, fish or meat? All living things need some kind of food to stay alive. Food gives them energy and helps them to grow and stay healthy. Green plants can make their own food. Some animals eat plants, and some animals eat other animals. Food energy passes from plants to animals in a long chain. This is called a **food chain**.

From plants to animals

All food chains begin with plants. Green plants take in water and a gas called carbon dioxide. Light from the Sun helps the plants join these together to make sugars. These sugars are the plants' food.

Next in the food chain come the animals that eat plants, such as sheep or rabbits. These animals are called **herbivores**. Herbivores are eaten in turn by animals that eat meat, such as foxes which catch and eat rabbits. Meat-eaters are called **carnivores**. Human beings and some other animals are called **omnivores**. They eat both plants and animals.

Find out more by looking at pages **34–35**

Trees and plants provide food, such as nuts and berries. Mice and squirrels eat nuts and berries.

Rabbits and mice feed on grass and seeds.

Foxes and hawks hunt and eat the rabbits, squirrels and mice.

More herbivores than carnivores

The grasslands of Africa are covered in grasses and other plants. There is always plenty of sunlight, carbon dioxide and water for these plants to grow and make their own food. Herds of antelopes feed on the plants. But a herbivore such as an antelope only uses about 10 per cent of the energy in its food. About 90 per cent of the energy is lost.

Small groups of lions prey on the antelopes. When a carnivore such as a lion eats another animal, only about 10 per cent of the energy in the food is used. The farther along the food chain you go, the less food there is available. This means there must be fewer lions than antelopes.

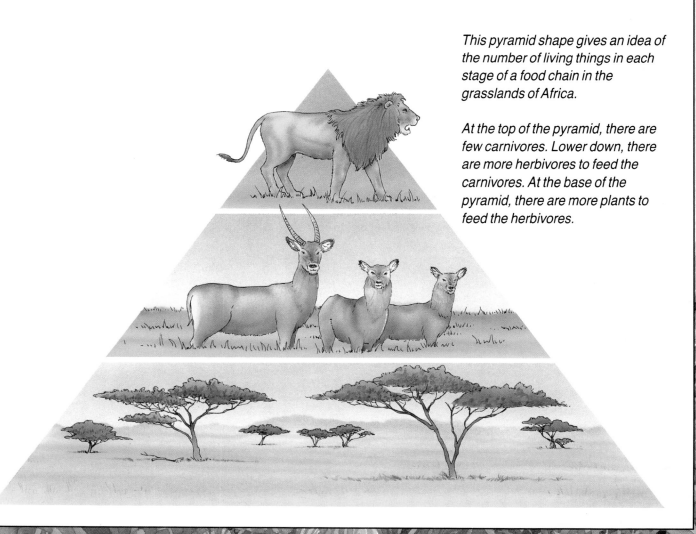

This pyramid shape gives an idea of the number of living things in each stage of a food chain in the grasslands of Africa.

At the top of the pyramid, there are few carnivores. Lower down, there are more herbivores to feed the carnivores. At the base of the pyramid, there are more plants to feed the herbivores.

Find out more by looking at pages **34–35**

Food chains in the sea

There are plants in the seas and oceans as well as on land. Tiny plants are part of a layer of millions of living things near the surface of the oceans. All these things are called **plankton**. Some plankton are plants that use light from the Sun to help make their own food. Other plankton are animals that eat the plants or each other.

Fish, shellfish and seabirds feed on the plankton. Then these creatures are eaten by larger animals, such as seals and some types of whale. This makes a food chain in the sea. So the food chains of the sea begin with plankton.

Do you sometimes eat fish? If you do, then you are at the end of a food chain.

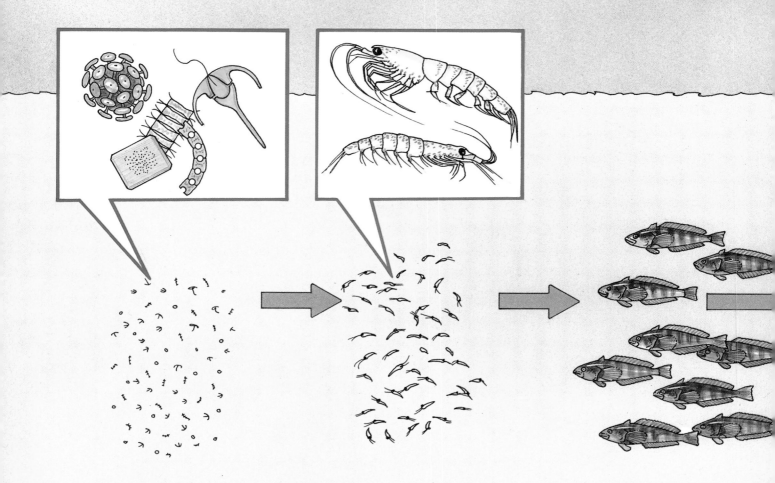

1. Microscopic plankton live near the surface of the sea.

2. Krill feed on the plentiful supplies of plankton.

3. Fish eat krill and other small creatures.

Antarctic food chain

Take a look at this food chain in the coastal waters of
Antarctica. Plant plankton are eaten by small crustaceans
called krill. The krill is eaten by a fish. The fish is eaten by a
seal. The seal is eaten by a whale. The whale is caught by
people, who sometimes eat the whale's meat. Oil from the
whale's body may be used to make products such as
cosmetics and soap.

Can you draw a food chain, starting with plankton and ending
with a human being?

4. Fur seals eat several kilograms of fish each day.

5. Killer whales sometimes eat seals.

6. People hunt and kill whales for food and oils.

Return to the soil

Find out more by looking at
pages **22–23**
26–27
32–33

energy from the Sun

All the food in the world initially comes from plants. They provide food for animals. These are then eaten by other animals which are in a food chain. So all animal life depends on plants.

Plants use carbon dioxide, water and light to make their food. Plants also soak up substances called **nutrients** from the soil, which make them grow. Nutrients include chemicals that contain nitrogen, which all plants and animals need. When an animal eats a plant, the nutrients in the plant pass into its body. If that animal is then eaten by another animal, the nutrients will be passed on again.

After the nutrients have been used by plants and animals, they return to the soil to help more plants to grow. The nutrients return to the soil through animals' droppings and through dead plants and animals that decay in the ground.

plant food

nutrients

bacteria

Bacteria at work

What makes dead plants and animals decay? They decay because millions of tiny **bacteria** feed on them. You may have seen a rotten apple or the remains of a dead animal lying on the ground. In time, bacteria will eat away at them and break them down into smaller and simpler pieces. In this way, bacteria break down dead plants and animals until they become part of the soil.

Bacteria use only a certain amount of the nutrients from the dead plants and animals that they eat. They return the rest of the nutrients to the soil. This is how the soil is provided with a fresh supply of nutrients over and over again, so that plants can keep on growing.

Bacteria feed on the flesh of this damaged fruit. The bacteria eat away at the apple until it breaks down and falls into the soil.

Energy from the Sun passes, in the form of food, from one living thing to another.

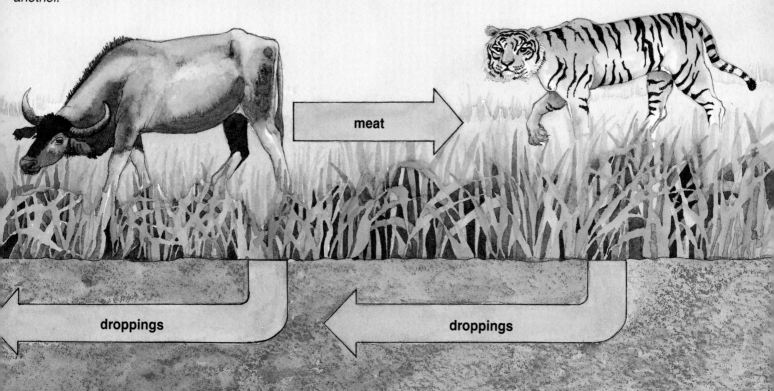

meat

droppings

droppings

These fungi are growing on a rotting tree. They help to break down the wood into small pieces.

What happens in the soil?

Have you ever seen toadstools growing on dead trees? Or a greenish mould on an old piece of bread? If you have, then you have seen **fungi** growing. Both toadstools and moulds are fungi. Fungi help to break dead plants and animals down into smaller and simpler pieces. This is part of the process of decay that happens to all dead things. When dead things decay, they are said to be decomposing. Living things that make dead plants and animals decay are called **decomposers.**

Fungi are important decomposers. Some fungi are so small that you can't see them without a microscope. These tiny fungi, along with bacteria, do most of the work in breaking down dead plants and animals and other waste.

Find out more by looking at pages **30–31**

Animal decomposers

Some animals help to decompose dead things. Many ants grow fungi on dead leaves for food. Dung beetles eat animal droppings, which are sometimes called dung. Other insects, such as woodlice and springtails, feed on decaying plants and meat.

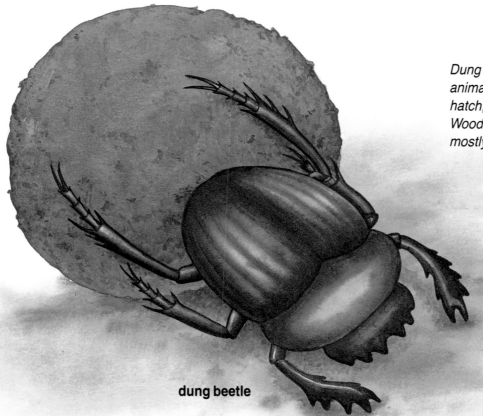

dung beetle

Dung beetles lay their eggs in animal droppings. When the eggs hatch, the young feed on the dung. Woodlice and springtails feed mostly on rotting plants.

springtail

What is humus?

When dead plants and animals in the ground have completely decayed, they form a dark-brown substance, called **humus**, in the soil. Humus holds water in the soil and provides the **nutrients** that plants need to grow.

So decomposers have an important job to do. Without them there would be no humus or nutrients in the soil and plants could not grow. The animals would die because there would be no food for them. The decomposers are the living things that keep life going on and on.

woodlouse

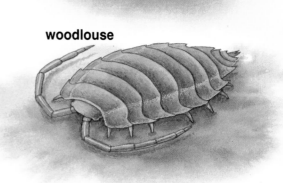

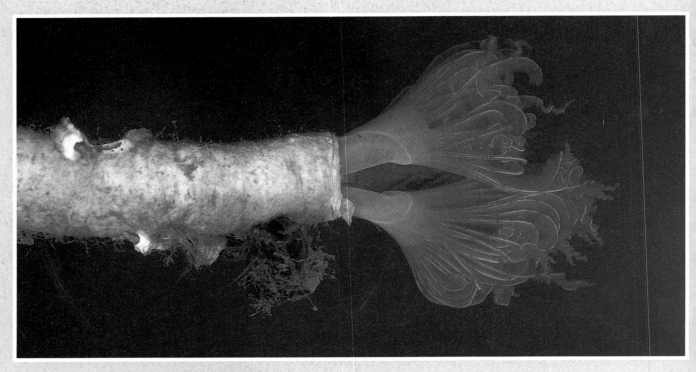

A fanworm catches food particles by waving its feathery tentacles around in the water.

Pelicans use their pouch to scoop fish out of the water. Then they swallow the fish whole.

How do living things eat?

Eating is easy! Most animals take food in at one end of their body. The food is squeezed through a long tube inside them, and the waste passes out at the other end of the body.

Many animals chew their food before it goes down into their stomach. Have you ever noticed a cow eating? Its jaws move round and round, grinding its food. Carnivores, such as tigers, have sharp teeth that tear their food with an up-and-down movement. Some animals don't chew their food at all. Many birds swallow fish or grain whole.

Different ways of eating

Other living things eat in very different ways. The **starfish** has a special way of taking food into its stomach. Its mouth is in the middle of its body. When the starfish is about to eat, it pushes its stomach through its mouth. The stomach then surrounds the food and digests it slowly, before returning to the body through the mouth. Waste food is also passed back out through the mouth.

Barnacles fasten themselves to rocks and other objects in the sea. They feed by sweeping their feathery tentacles through the water. They are called **filter feeders** because they use their tentacles like a sieve to pick up tiny particles of food floating close by. Many other sea animals, including fanworms and corals, are filter feeders.

Unlike most animals, **termites** can eat dry, dead wood. They can digest the wood because they have special bacteria in their gut.

Earthworms feed on tiny particles of dead plants in the soil. As they move along, earthworms swallow the soil and digest the particles of dead and decaying plants. The rest of the soil passes through their body and is left behind as waste.

Aphids feed on plant juices. They have a mouth shaped like a tube, which can pierce the plant stems or leaves. The aphids suck the plant juices rather like people suck a drink through a drinking straw.

How do green plants eat?

Green plants make their own food by a process called **photosynthesis**. They use light from the Sun to combine a gas from the air, called carbon dioxide, with water from the soil. With this compound, the plants make the sugars which they can change into energy.

Find out more by looking at pages **18–19**
36–37
40–41

Aphids pierce the leaves of plants with their sharp mouthpiece. Then they suck up the plant juices.

Find out more by looking at pages **38–39**

Using food

What happens to food once it's inside an animal's body? Food contains many different substances, such as **proteins** and **carbohydrates**, that living creatures need. Carbohydrates are needed to provide energy, and proteins are needed to help animals grow and stay healthy. The food has to be broken down into lots of smaller, simpler substances, so that the body can absorb it. This process is called **digestion**.

Juices at work

Digestion begins in the mouth when an animal chews its food. A digestive juice, called saliva, starts breaking down the food. When the food reaches the stomach, it's broken down some more by acid. Here, chemical substances called **enzymes** also break down the food.

After the acid and enzymes break down the food in the stomach, the food is slowly pushed along into a tube called the **small intestine**. Here, the food is finally digested. All the important parts of the food — the nutrients — are absorbed into the wall of the small intestine. They then enter the blood and are pumped round the body.

The stomach of a cow or a sheep has four sections. The grass that the animal eats is chewed more than once before it reaches the intestine.

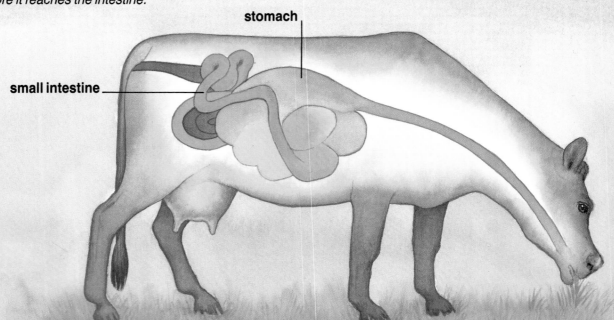

stomach

small intestine

A special stomach

Some plant-eaters, such as goats and cows, have a very interesting digestive system. They have a stomach with four separate sections. When the animal swallows a mouthful of grass, it is collected in the first and second sections. In the second section, the grass is softened into the **cud**. As the animal rests, stomach muscles return the cud to its mouth where it is chewed again. After the second chewing, the food goes through all the sections of the stomach and then to the intestine.

Some spiders digest their food outside their body. They do this by injecting enzymes into the insects they catch. After a while, the enzymes break down the insects. Then the spiders suck the digested insects into their body. Houseflies digest their food by squirting enzymes over their food and sucking up the liquid that results.

Why do animals need food?

Animals need food because it gives them energy. Energy is what keeps their body moving and working properly. As well as providing energy, food makes bodies grow and also repairs parts that have become worn out. After the food is digested, the body uses the chemicals from the broken-down food to build up the body. Animals can also store some types of food as fat inside their body. It is used whenever it is needed.

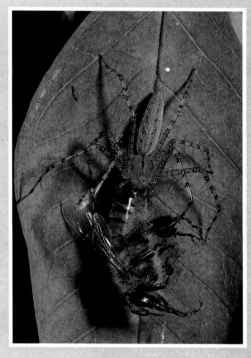

The lynx spider's prey is often much larger than itself. The spider injects enzymes into the insect's body. These turn the body to a juice that the spider can digest.

The leopard digests its food in much the same way as we digest our food.

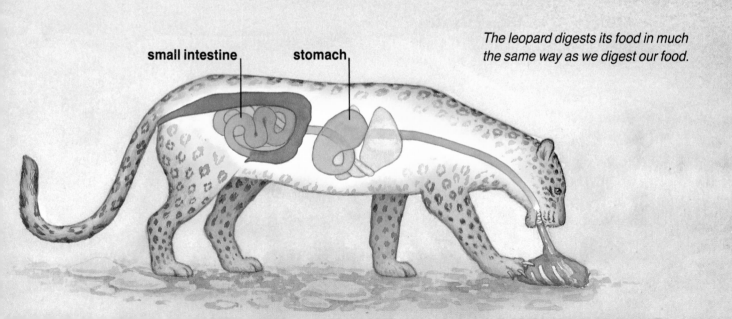

small intestine   stomach

Find out more by looking at pages **36–37**
40–41

Getting rid of waste

Food keeps living things alive. But animals cannot use all the food they eat. As food passes through the body, only some of it is digested. The body takes in all the parts of the food that it needs and then gets rid of the rest. The waste food is sent through the lower part of the body and finally comes out of the opening called the **anus**. This solid waste food that comes out of the bodies of humans and other animals is called **faeces**.

As well as solid waste food, animals also need to get rid of other substances that are harmful to their body. In vertebrates, this job is done by the kidneys. The kidneys collect all the unwanted chemicals and excess water and turn them into a liquid called **urine**. The urine is then stored in another part of the body, called the **bladder**, until it leaves the body.

Birds produce faeces and liquid waste together through one opening. Their droppings are faeces in the middle with a coat of thick, white liquid waste on the outside. Other kinds of animal, such as reptiles and insects, also release their waste food through one opening. Some simple animals don't have a special opening for waste. Their waste products have to go out through the mouth.

Mammals release carbon dioxide, urine and faeces from their body. If for any reason an animal is not able to get rid of this waste, it will become ill.

Plant waste

Even plants clear out the things they don't need. Some mangrove trees, for example, have too much salt in them because they grow in salty places. They get rid of the salt by pumping it into special leaves. These leaves then fall from the tree.

Mangrove trees can get rid of waste by storing it in special leaves. These leaves fall from the tree, taking the waste with them.

Other types of waste

What else do animals take into their bodies, use and then send out as waste? The answer is gases. We all breathe in a gas called **oxygen** from the air. Oxygen helps the food inside our body to produce energy. This energy production inside us makes a waste gas called **carbon dioxide**. We get rid of this by breathing it out into the air.

Breathing

We need to breathe to stay alive. When we breathe, we take in a gas called **oxygen** from the air. Almost all living things need oxygen to survive.

When an animal breathes, the oxygen it takes in enters its blood and is transported to every cell in its body. The food it eats contains sugars which are also transported to every cell in the body. These sugars are split up in each cell by chemicals called **enzymes** and energy is released. Oxygen is essential to this process of splitting up sugars and other foods, and releasing the energy stored within them.

When a cell releases energy from food, a gas called **carbon dioxide** is released. This gas enters the blood and is breathed out at the same place as oxygen is breathed in.

How different kinds of animal breathe

All **mammals**, **birds** and **reptiles** breathe by taking air into parts of their body called **lungs**. Lungs are like stretchy bags that grow larger as air flows in. They become smaller as air and carbon dioxide are breathed out.

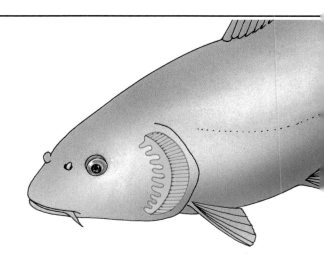

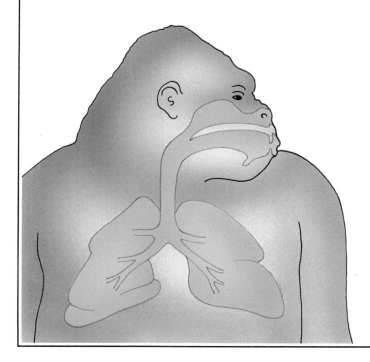

Fish and many other water animals take in water and pass it through parts called **gills**. They absorb the oxygen that is dissolved in the water.

Mudskippers are fish that come out onto land to feed. They can't breathe air, so they take a mouthful of water with them when they leave the sea. This keeps their gills wet while they are out of the water. When they have used all the oxygen from their mouthful of water, they take some more water from a puddle.

Plants breathe, too!

Even plants need to take in oxygen from the air. The oxygen helps a plant's cells to release energy from sugars that are stored inside the plant. Plants give out a small amount of carbon dioxide, especially during the night.

During the day, when plants make their food, they do the opposite. They take in carbon dioxide from the air and give out oxygen. They do this in sunlight by **photosynthesis**. The oxygen and carbon dioxide pass in and out of the leaf through tiny holes called **stomata**. Each stoma has two guard cells that help it to open and close. Plants give out oxygen, which animals breathe in, and animals breathe out carbon dioxide, which plants take in.

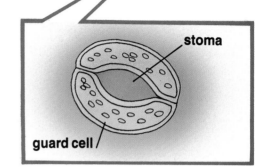

stoma

guard cell

Insects breathe through a network of tubes. These tubes reach every part of the insect's body.

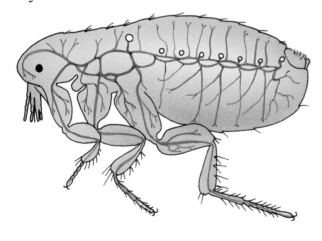

Water spiders breathe air, so they take their own air bubble down into the water.

Energy for living

Do you like to play games that make you run very fast or jump in the air? You need a lot of energy to do these things. Energy is what makes our bodies move around. Every time you walk, ride a bicycle or move in any way, you are using up energy in your body. Energy also makes substances inside your body move around so that your body can work properly. Without energy, your heart would stop beating and you would soon die. So people need energy to live. In the same way, all living things need energy so that their body can stay alive and work properly.

Different animals have very different energy needs. This chart shows the amount of food each creature eats in a week compared with its body weight.

**puma
44 kilograms**

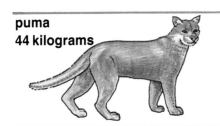

A puma weighing 44 kilograms eats about three and a half kilograms of food each day. That's eight per cent of its body weight.

a week's food

24.5 kilograms

**locust
1 gram**

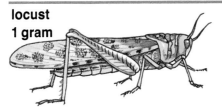

A locust weighing one gram eats about half a gram of food every day. That's 50 per cent of its body weight.

**penguin
40 kilograms**

A penguin weighing 40 kilograms eats up to 80 kilograms of food every day. That's twice its body weight.

**African elephant
5,400
kilograms**

An African elephant weighing 5,400 kilograms eats about 349 kilograms of food every day. That's more than six per cent of its body weight.

2,443 kilograms

**Etruscan shrew
2 grams**

An Etruscan shrew weighing two grams eats about six grams of food every day. That's three times its body weight.

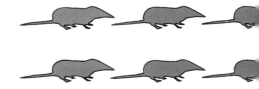

How do living things get energy?

Living things get energy from their food. Food contains **sugars**, which are chemicals that contain stored energy. As food is digested and broken down into simple substances, these sugars join with oxygen that is breathed in from the air. This process releases the energy in the sugars.

The sugars contain a substance called **carbon**, which is needed to make all living things. The energy is released when the oxygen that is breathed in joins with the carbon. The joining of carbon and oxygen also produces the gas carbon dioxide. This is why living things breathe out carbon dioxide into the air.

Find out more by looking at pages **36–37**
40–41

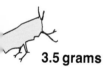
3.5 grams

560 kilograms

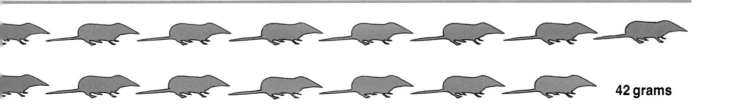
42 grams

Essential supplies

Did you know that parts of the food you eat are carried to your fingers and toes? In fact, they are carried all round your body. The different parts of your body are made up of lots of tiny cells. All cells need a supply of food to keep them alive and well. They also need a supply of the oxygen you breathe in from the air.

This is true of most animals and plants. All the cells inside them need a supply of food and oxygen. But how is the food and oxygen carried round all the different parts of the animals and plants?

Blood makes the delivery

In many animals, food and oxygen are carried round the body by the **blood**. The blood is pumped round and round the body by the heart. As the blood travels, it picks up oxygen from the lungs and digested food from the stomach and intestine and from a special store called the liver. It then carries the oxygen and food to all the cells as it moves round the body.

Cells are living things and, like other living things, they produce waste. The waste substances are carried away by the blood as it moves round the body again.

In mammals, blood travels round the body inside tubes called **blood vessels**. Birds, reptiles and fish also have blood vessels. But the blood of insects flows openly round their body instead of in tubes. The blood fills up the insides of their body and supplies the cells with food. But it does not take oxygen to the cells like blood does in other animals. Insects take in oxygen through tiny holes in the sides of their body.

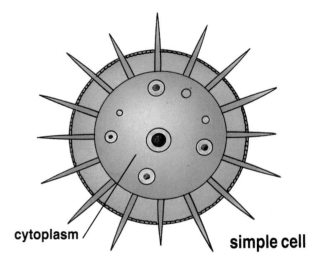

cytoplasm / **simple cell**

In a single-celled organism, food is carried round by the moving cytoplasm.

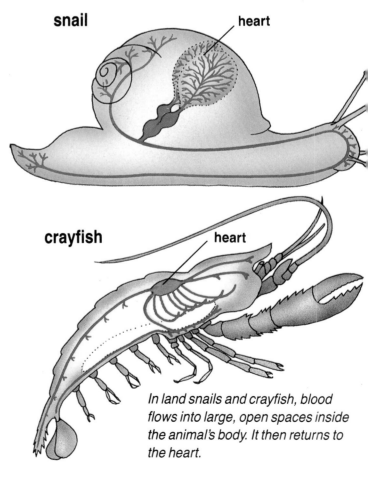

snail **heart**

crayfish **heart**

In land snails and crayfish, blood flows into large, open spaces inside the animal's body. It then returns to the heart.

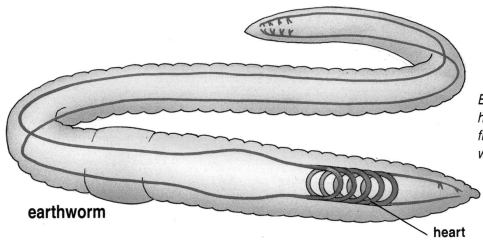

earthworm

heart

Earthworms have five pairs of hearts! Blood vessels branch out from the hearts to supply the body with blood.

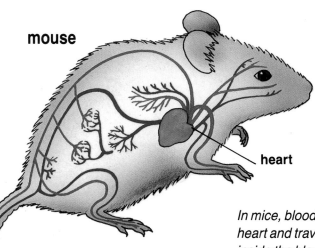

mouse

heart

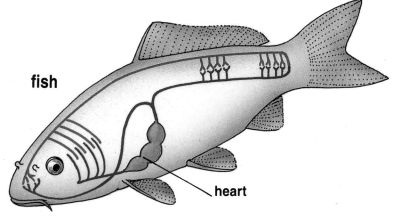

fish

heart

In mice, blood is pumped from the heart and travels through the body inside the blood vessels. In fish, blood is pumped from the heart to the gills, before travelling to other parts of the body.

What is plant sap?

In plants, food has to travel from the leaves, where it is made, to other parts, such as the stems and roots. Plants have a juice inside them called **sap**. The sap is like the blood of the plant. It flows through special tubes, called **phloem**, which carry the sugars to all the plant's cells.

Another kind of sap containing water and minerals is carried by special vessels, called **xylem**, from the roots of a plant up to its leaves.

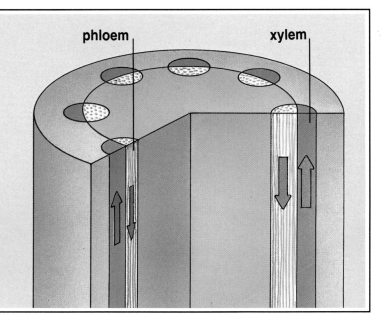

phloem   xylem

Moving about

You need your muscles to make your body move. Other animals also need their muscles, so that they can catch their food, protect themselves and do many other things. Muscles can make bodies run, walk, turn and bend.

Some of the muscles in your body are joined to bones. They make the bones move at places called joints. The muscles do this by becoming shorter and fatter. This makes them pull at the bones so that they move. Muscles can only pull – they can't push. This means that muscles of this kind have to work in pairs. One pulls the bone one way and the other pulls the bone back again. In your arm, your **biceps** makes your elbow bend and a muscle called the **triceps** pulls your arm straight.

When an insect flies, its wings beat up and down. These wing movements are controlled by different muscles in the insect's body. In turn, these muscles become tight or relaxed.

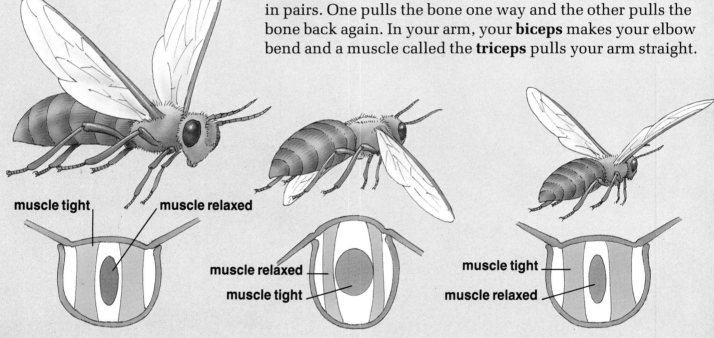

muscle tight   muscle relaxed

muscle relaxed
muscle tight

muscle tight
muscle relaxed

Feel your muscles!

1. Hold up your arm as shown in the picture.

2. Place your right hand on the upper part of your left arm.

3. Now clench your left fist against your left ear. You will feel the muscle moving. This muscle is the biceps.

Bones help, too!

Human beings and many other animals have their bones inside their body. But some animals have their bones on the outside. You may have seen a beetle crawling up a grass stem. Inside the beetle's legs there are tiny muscles just like yours, which pull the animal's legs backwards and forwards. You can't see its muscles moving because they are inside a hard outer casing called the **exoskeleton**. The beetle's muscles are firmly attached to the inside of its exoskeleton.

Not all animals have both hard bones and muscles to make them move. Earthworms do not have bones. They use their muscles to crawl. The earthworm stretches the front part of its body and then pulls up the other part to move along.

Watching animals move

If you watch small animals walk across a sheet of glass or plastic, you can see exactly how they move.

You will need:

a sheet of clear glass or plastic

some small animals, such as woodlice, earthworms, snails and beetles

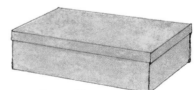

a cardboard box

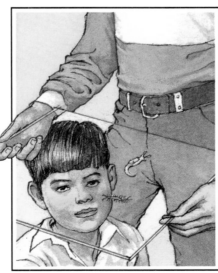

1. Ask an adult to help you find a sheet of glass or plastic and hold it safely for you.

Be careful that the animals you have chosen are not poisonous.

2. Place the animals on the glass or plastic one at a time, and then look at them from below. This way, you can see their movements very clearly.

Keep the animals in a dark, cool box while not in use. When you have finished with the animals, take them back to the place where you found them.

Keeping in touch

When you turn the pages of this book, you are moving your hands and fingers. As you read, your eyes are moving as they follow the words on the page. These body movements are happening because you are making them happen. You could easily stop them if you wanted to. This kind of body movement is called a **voluntary movement.** Voluntary movements of the body are all the movements that you make your body do, such as walking, running or picking up a pencil.

Your body also moves in another way. While you are reading this book, your heart is moving as it beats. Parts of your stomach and gut are moving as they digest your food. You do not make these body movements. They happen anyway. This kind of body movement is called an **involuntary movement.**

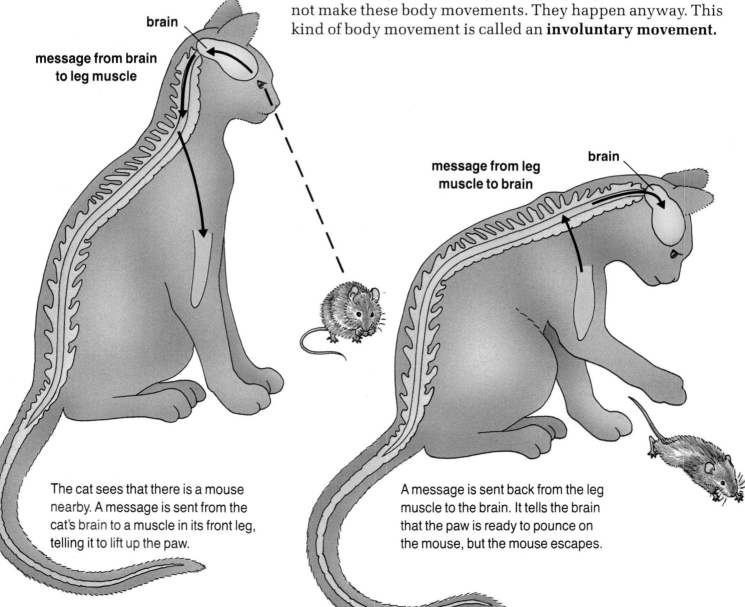

brain

message from brain
to leg muscle

message from leg
muscle to brain

brain

The cat sees that there is a mouse nearby. A message is sent from the cat's brain to a muscle in its front leg, telling it to lift up the paw.

A message is sent back from the leg muscle to the brain. It tells the brain that the paw is ready to pounce on the mouse, but the mouse escapes.

What does the nervous system do?

How do the parts of your body know when to move? The answer is that your brain is sending them messages through a network of nerve cells that make up the **nervous system**. When you decide to pick up a pencil, your brain sends a message through the nervous system to the muscles in your arm and hand to start moving. The nervous system also carries messages to the parts of your body that move without your control, such as your heart and stomach.

Most animals have some kind of nervous system. But no animal has a brain as well developed as the one in human beings. When animals are in danger, messages travel through the nerve cells to tell the different parts of their body to do something at once. If a mouse sees a cat, messages will be sent quickly through the nerve cells of the mouse to tell it to run away. In the same way, a snail will pull its head into its shell for protection when it senses a threat.

Chemical messages

There is another way that messages are passed round the bodies of human beings and other animals. This is by using chemical substances called **hormones**. Most hormones are made in organs of the body called **glands**. These glands are found in various parts of the bodies of human beings and most animals. An important gland is the **pituitary gland**. This is often called the master gland. It lies just beneath the brain and produces many different hormones.

Hormones help to control digestion, growth and other body functions. When you chew food in your mouth, your hormones carry messages from your mouth to your stomach and gut, telling them to get ready to digest the food.

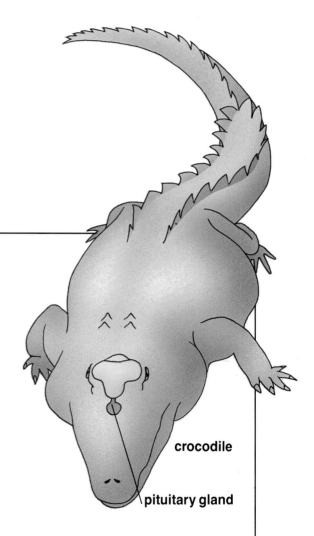

crocodile

pituitary gland

The pituitary gland lies just beneath the brain. Although it is only the size of a pea, it produces many different hormones and is one of the most important glands.

Find out more by looking at pages **18–19**

What is reproduction?

Human beings have babies. Fish lay eggs which hatch into young fish. Trees grow seeds which can grow into new trees. All living things make copies of themselves. This is the process of **reproduction**.

Asexual reproduction

There are two different kinds of reproduction. In one kind, there is only one adult of the species, and the young ones look exactly the same as their parent. This kind of reproduction is called **asexual reproduction**. Living things reproduce like this in several different ways.

The simplest living creatures reproduce by just **splitting** in two. Most creatures that are made up of only one cell reproduce in this way. Even some simple animals such as sea anemones, whose bodies are made up of many cells, can do this

Many plants can make new plants from a small piece that has broken off, such as a twig or stem. This is useful if the plant has been damaged by animals or by strong winds. It is also useful to gardeners, because it means that they can take **cuttings**, which grow into new plants. The plants grown from cuttings will be exactly the same as the parent plant.

The underground stem of the cowslip sometimes develops new stems. After a few years, the oldest part of the stem dies, and the plants are separated.

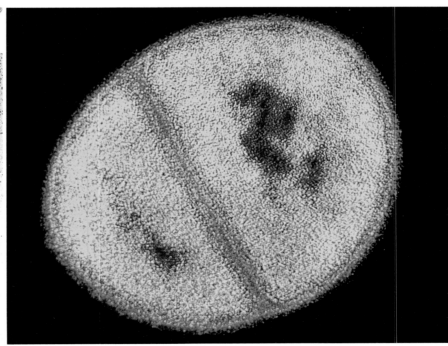

This bacterium is in the process of dividing into two. It does this by itself, without the help of another bacterium.

The male bullhead fish guards eggs from predators.

Sexual reproduction

In the other kind of reproduction, called **sexual reproduction**, there are two adults of the species — one male and one female. The young ones are not quite the same as the parents. Human beings reproduce in this way. You don't look exactly like your parents, although you may look a bit like both of them. The same is true of other animals and plants that have two parents. They take some of their features from one parent and some from the other.

For sexual reproduction to take place, one cell from the male must join with one cell from the female. In animals, the male produces cells called **sperm** and the female produces **egg** cells. When one egg cell joins with one sperm cell, the egg cell begins to grow into a new animal. This process is called **fertilization**.

Fertilization can happen inside or outside the female's body. In most fish and frogs, the female lays eggs and the male releases his sperm over them. In birds and mammals, the male's sperm fertilizes the eggs inside the female's body.

In plants, the male cells are called **pollen**. Pollen is carried from male flowers to female flowers by insects or by wind. The pollen joins with the egg cells and produces seeds, which grow into new plants.

Find out more by looking at
pages **22–23**
        24–25
        34–35

What is a parasite?

All animals feed on other animals or plants or both. But some small animals actually live on, or even in, the animals or plants that they feed from. They eat small amounts of the living bodies of larger animals or plants, which also provide shelter for them. Animals that do this are called **parasites**. Some plants are parasites, too. They grow on other plants and feed from them. They don't make food for themselves as other plants do. The animal or plant that parasites live on is called the **host**.

Animal parasites

Some animal parasites feed by sucking the blood of larger animals. These blood-suckers include **ticks**, **fleas**, **lice** and **leeches**. They cling to the body of the host animal and bite into its skin. Their body becomes swollen as it fills with blood. Fleas jump from one part of the host animal to another, sucking its blood.

Fleas live in clothes, and on the bodies of humans, birds and other animals. They suck the blood of a host animal.

Plant parasites can do a lot of damage to crops. This brown rust is growing on a barley leaf.

Dodder stems wrap themselves around a host plant. Special suckers become attached to the host plant, and share the host's food supply.

Plant parasites

Many plants that are parasites, like **dodder** or **rafflesia**, don't have green leaves. And they don't have green stems like other plants. Other plants are green because they contain a green substance called chlorophyll. This absorbs the sunlight they need to make their food. Plants that grow and feed on other plants don't make their own food, so they don't have green chlorophyll.

Fungi

Some kinds of **fungus** are parasites. Fungi can grow on both plants and animals and can cause serious diseases. The fungi that grow on plants can sometimes spread very quickly and destroy important crops, such as wheat and potatoes.

The smallest parasites

Some bacteria are parasites that can cause disease. They make animals ill when they invade their bodies. Even smaller than bacteria are parasites called **viruses**, which can also cause disease. The body of a virus has no mouth or stomach. The simplest viruses don't even have an outer skin. And they have none of the jelly-like substance called **cytoplasm** that is in all other living things.

Defences

Most animals and plants live in danger of being killed or harmed by other living things. So they have to have some way of defending themselves. A species of plant or animal with no protection would not survive.

Inner protection

Both animals and plants can defend themselves against harmful bacteria and viruses. Animals have special cells that fight and kill these bacteria and viruses. These cells are always ready to protect any part of the body that is attacked.

When an animal's skin is pierced and bacteria enter the wound, **white blood cells** quickly crowd into the hurt area. The cells break down the harmful bacteria by surrounding and eating them. The broken skin is gradually repaired by new skin cells that grow across the wound.

Plant cells have strong walls to keep out bacteria and viruses. But when insects feed on plants they break down the cell walls. Bacteria and viruses can then enter the plant and cause disease.

Aphids not only damage the plant they feed on, but can also spread viruses and bacteria from plant to plant.

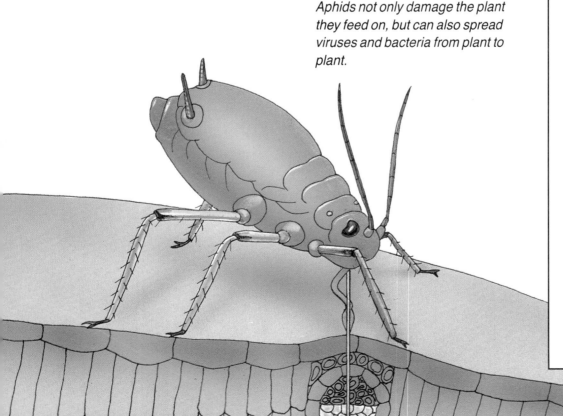

Fighting infection

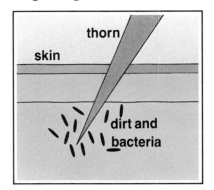

1. A thorn digs into the skin of an animal, letting in dirt and bacteria.

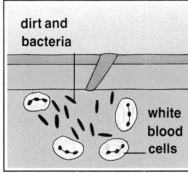

2. White blood cells cluster around the wound.

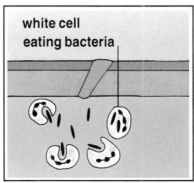

3. Some white cells destroy the bacteria by eating them.

55

Find out more by looking at
pages **24–25**
34–35

*The spines of cactus plants provide
good protection from animals which
might want to eat the plants.*

Outer protection

Insects need good ways of defending themselves because
there are so many larger animals that can eat them. Bees and
wasps can inject poison into their enemies with their sting.
Many ants have a poisonous bite or sting. Plants cannot move
so they need special protection, such as prickles, hairs or
tough skins.

Warning colours

Poisonous insects are often brightly coloured. Red and black
or yellow and black are the most common colours found on
these creatures. To other animals these colours mean 'stay
away!' They are called **warning colours**.

Camouflage

Many animals defend themselves by blending in with their
surroundings. This is called **camouflage**. When they keep
still, some insects look like twigs, leaves or flowers.

This fossil of a reptile was found in Switzerland. The animal probably lived about 200 million years ago.

Then and now

Rocks often contain the remains of living things that died long ago. These remains are called **fossils**. The kinds of rock that contain most fossils are called **sedimentary rocks**. Sedimentary rocks build up in layers very slowly. So we know that the rocks near the surface, just under the soil, are younger than the layers of rock below them. The oldest rocks are right at the bottom. By studying chemicals in the rocks, scientists can tell how old they are.

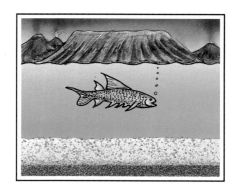

1. This prehistoric fish probably lived in the sea about 400 million years ago.

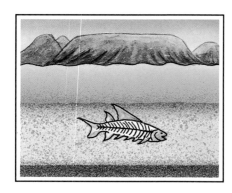

2. When the fish died, its body sank down to the sea-bed.

3. Sand and mud covered up the fish's body. The flesh decayed and the body became a fossil.

Secrets in the rocks

Rocks that were made less than 300 million years ago contain fossils of all kinds of different animals and plants, including horses, tigers, birds, dinosaurs, dragonflies and flowers. In rocks over 300 million years old, fewer animals and plants have been found, although there are plenty of fossils of fish, plants without flowers, and simple animals such as molluscs.

Older rocks, made about 500 million years ago, contain no vertebrate fossils. In even older rocks, going back to about 600 million years ago, there are few animal fossils. The only animal ones are fossils of very simple creatures like jellyfish and worms. The rocks made 3,000 million years ago contain no fossils of animals at all, only microscopic fossils that are probably bacteria.

Scientists believe that **fossil records** show that life began with very small, simple creatures. As time went by, more and more complex creatures appeared. The most complex animals and plants — mammals, birds and flowering plants — appeared last of all.

This trilobite looks similar to a modern woodlouse. Trilobites lived in the sea.

Find out more by looking at pages **56–57**

Survival

Some scientists believe that living things have developed from other simpler living things, which have slowly become more complex over millions of years. There are enough fossils to show clearly how some kinds of animals have developed. Studying fossils has helped scientists to reach the conclusion that mammals have the same ancestors as reptiles. Other fossils show what the ancestors of some animals, such as the camel, looked like. Camels developed from a small animal the size of a fox into larger animals with hooves, a long neck and large teeth.

As animals have developed, they have become better suited to their surroundings. The giraffe's long neck, for example, enables it to reach high up into trees and eat the leaves there. Scientists believe that animals have developed because their ancestors were the ones that could make good use of their surroundings and so were able to survive.

Some animals, such as the dinosaurs, had no chance to develop. No one knows exactly why they died out. Other species have disappeared completely. Some were hunted until they became extinct. Others were destroyed by disease or by changes in climate.

Over millions of years, elephants became bigger and heavier. The mammoth was about four and a half metres tall. There are now only two kinds of elephant — the African and the Indian elephant. The African elephant is about three and a half metres tall.

moetherium **gomphotherium** **platybelodon** **mammoth** **African elephant**

How living things have developed

Living things have developed over millions and millions of years into their present forms. Scientists believe that the first living things appeared about 3,500 million years ago.

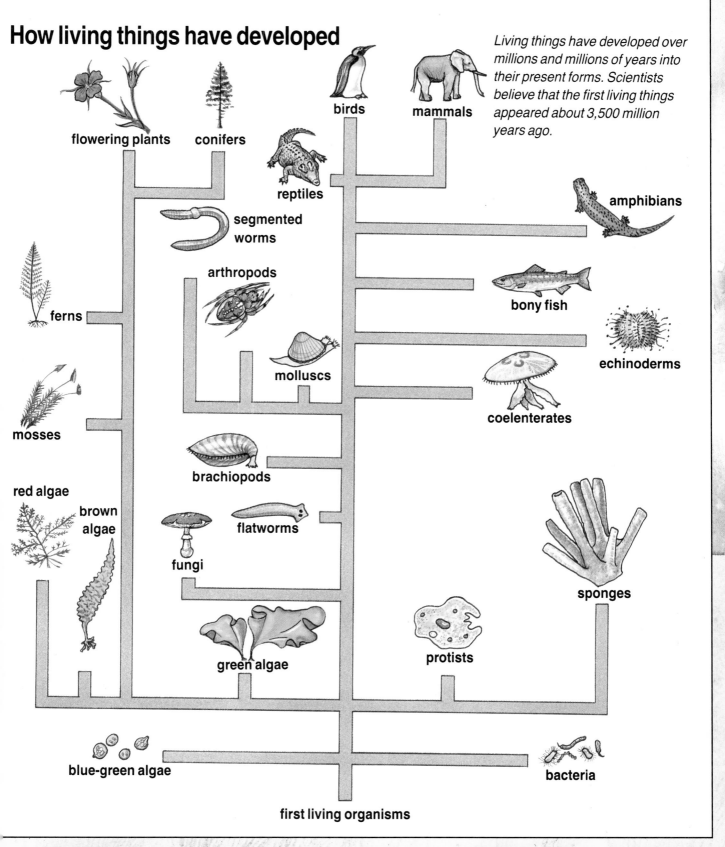

flowering plants

conifers

birds

mammals

reptiles

amphibians

segmented worms

arthropods

bony fish

ferns

molluscs

echinoderms

mosses

coelenterates

brachiopods

red algae

brown algae

fungi

flatworms

sponges

green algae

protists

blue-green algae

bacteria

first living organisms

Index